GUIDE TO THE GEOLOGY OF BARTOW COUNTY, GEORGIA

by

WM. HOVEY SMITH

BUDGET PUBLICATIONS
Whitehall
Rt. 1 Box 603
Sandersville, Georgia 31082

With the cooperation of
THE WILLIAM WEINMAN MINERAL CENTER AND MUSEUM
P.O. Box 1255
Cartersville, Georgia 30120

GUIDE TO THE GEOLOGY OF BARTOW COUNTY, GEORGIA

by

Wm. Hovey Smith

TABLE OF CONTENTS

TABLE OF CONTENTS

page

ILLUSTRATIONS

TABLES

INTRODUCTION

ROCKS AND MINERALS and how and why they came to be seen in Bartow County as they are today is what this book is about. For some, this information will be an adequate overview of the county's geology, but others will find that this sparse historical treatment will raise more than a few questions.

The annual value of county's mineral production usually places it among the the top five of Georgia's 159 counties. At this time there are four active mining companies which produced $4,375,000 worth of minerals during 1975, the last year for which results were reported.

In the past there have been several dozen companies mining a variety of minerals. Not only do these minerals include the barite, ocher, umber, ceramic shales and sand and gravel that is mined now; but important amounts of brown iron ores, manganese ores, limestone, gold, cement rock, graphite, dolomite, aluminum ore, slate, building stone and brick clay have also been extracted from beneath the county's red soils.

Hundreds of thousands of man hours expended by scores of geologists have resulted in a significant body of published information. This literature has been freely used, and references to many of these works are included.

During the Civil War, World War I, World War II, and the Korean Conflict, the mineral resources of Bartow County have been called upon to supply vitally needed raw materials. As a storehouse with remaining reserves of barite, iron and manganese ores, the county's minerals remain important to the nation.

Thanks are extended to The William Weinman Mineral Center and Museum in Cartersville for sponsoring this work and to Stan D. Bearden of New Riverside Ochre Co. for providing geological and processing information.

Location

Bartow County is located in Northwest Georgia. Among the towns within the county are Cartersville, the county seat; Cassville, the antebellum county seat; Emerson; Kingston; Adairsville; Rydal and Euharlee.

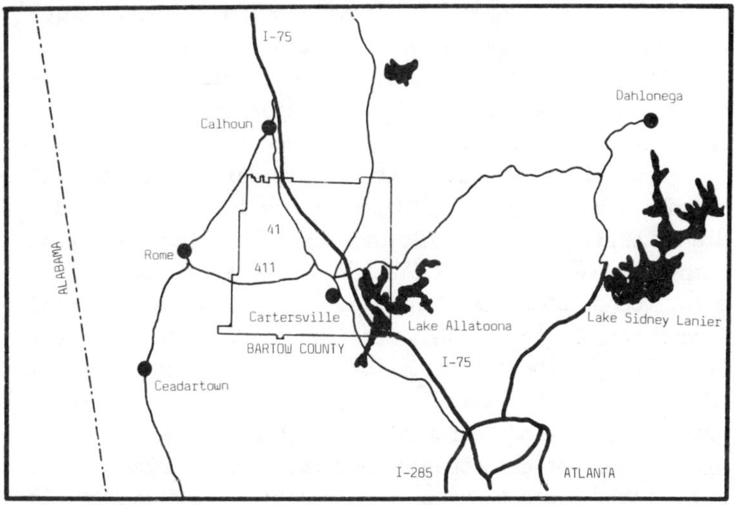

Figure 2. Location map.

It is served from its southeast corner to nearly its northwest corner by Interstate 75, while U.S. 411 passes through the northeast third of the county from north to south and then changes to an east-west route between Cartersville and Rome.

Bartow County ranks 38th land area and 24th population among Georgia's counties. It has a mean annual precipitation of 53 inches and 200 frost free days per year.

Mining History

America's Indians did the first mining in Bartow County when they settled the eastern part of the United

States some 10,000 years ago. They used flint weathering from the county's carbonate rocks to make their spear and arrow points, stones to make their axes and hoes and clay to daub their huts and to make their pots. They were as dependent on the earth to derive the things they needed to survive as we are today.

Examples of the Indian's use of the county's mineral resources are seen at the Etowah Indian Mounds, which were occupied between 1000 and 1650 A.D. The Mississippian Mound Builders used clay, flint, marble, manganese oxides, graphite, ochre and even native copper from Michigan for construction, utilitarian and ornamental uses.

The first white man to mine in Bartow County was probably an anonymous gold seeker who intruded into Indian territory after gold was discovered in White and Lumpkin counties in 1829 (Yeates and others, 1896). The gold found on Indian lands led to the Cherokees being forced to sell their lands for $5,000,000, their removal from Georgia and the formation of Cass County. Bartow County was first called Cass County which was, in turn, originally part of Cherokee County (Cunyus, 1933).

A vein of gold-bearing quartz, the Allatoona Vein, and nearby gold-bearing creek deposits were worked in 1835-1840, 1878 and in 1932, according to Kesler (1950). These and other properties, notably the Glade Mine, had a modest production compared to the more numerous deposits in the Dahlonega District of Lumpkin County. No production figures for Bartow County are known to be preserved, but Yeates (1896) states that the near-surface portions of the veins were reputed to be rich.

The Allatoona Vein and associated gold-bearing stream, or placer, deposits are now beneath Lake Allatoona.

Outside of clay, sand and stone which the early settlers used to build their houses, the next mineral to be extracted was brown, limonitic iron ore which was found to be abundant in southern Bartow County.

The first makers of iron in the county whose names are recorded were Moses and Aaron Stroup, two brothers

Figure 3. View of Hodge Mine located about two miles northwest of Euharlee. This water-filled pit was one of the largest mining operations in Bartow County. Iron ore was mined from this pit until 1958.

from Germany who began operations in 1837 (Cunyus, 1933). The same author reports that Mark A. Cooper purchased an interest in the Stroups' furnace on Stamp Creek and also made pig iron.

Cooper realized that he had to make finished articles rather than being just a supplier of raw iron if his business was to prosper. In 1845 he formed the Etowah Manufacturing and Mining Company. This aggressive enterprise installed a rolling mill, a 10-machine nail factory, a machine to make railroad spikes, a blast furnace and a foundry for making castings on the banks of the Etowah River near the present Allatoona Dam.

Charcoal made from nearby forests provided the fuel to reduce the iron ore to metallic iron; coal from Dade County was used to fire the forges for working it, and waterpower from the nearby river powered the factories' machines.

One of Gen. William T. Sherman's objectives on his 1864 march to the sea was to destroy Georgia's ability to make war materials. His men burned Cooper's factories, and all that remains of a thriving enterprise that employed hundreds of people during the early 1860s is the stack of Cooper's Furnace shown on the rear cover.

One of the periodic attempts to revive Georgia's iron industry was made by Georgia's Civil War governor, Joe E. Brown, who started a furnace in the 1870s at Rogers Station on the Western and Atlantic Railroad. This furnace closed in 1880 (Haseltine, 1924).

As it became more apparent that small iron making facilities could not compete with larger, more efficient, operations, Georgia's furnaces ceased operations. The last iron reduction furnace in Rome closed about 1910 (Haseltine, 1924).

Miners had to ship their ore further and further to market. The last large scale mining of iron ore in Bartow County was done by John W. Hodge who mechanized the local mining industry. He began operations in 1946 and ceased in 1958. The last ore he shipped to an Alabama smelter in 1963 was from a lot of ore that had been mined earlier and stockpiled. In all, about 18 million pounds of iron ore has been mined in Bartow County (Hurst and Crawford, 1970).

A vital mineral for making black powder, saltpeter, was mined in an unusual setting in Bartow County. This mineral is made of potassium and/or sodium nitrate and was recovered prior to and during the Civil War from the earth of the floor of Kingston Cave.

Saltpeter was eagerly sought by a beleaguered Confederacy in her fight for independence. It was recovered from the cave earth by washing it with water and boiling the water until the saltpeter crystals could be separated. The crystals were shipped to the Confederate powder works in Augusta where they were used to make black powder by combining them with charcoal and sulfur. Relics of the Civil War mining remain in the cave where they are now protected under Federal law.

Manganese, a vital element used in steel alloys, is associated with some of Bartow County's iron ores, and nearly pure masses of manganese oxides were sometimes encountered. The first known shipment of manganese ores from the county was 550 tons mined near Cartersville in 1866 (Watson, 1908).

According to Watson (1908), six companies were mining manganese in 1902. He estimated a total production of 45,000 tons between 1866 and 1888. During World War I production increased to meet the demand for the war effort, and McCallie (1926) reports that over 100,000 tons of manganese ores were mined from the Cartersville District during the First World War. Additional mining occurred during World War II and during the federal strategic metals stockpile program in the 1950s. According to statistics compiled by the U.S. Bureau of Mines, $1,041,813 worth of manganese ores, about 500,000 tons, had been mined from Bartow County through 1970.

No company in the county is now mining manganese, and remaining reserves of ore are considered to be small.

A yellow iron oxide pigment called ocher and a related tawny brown pigment named umber also attracted the attention of 19th Century miners. The first authenticated record of ocher being mined in Bartow County was by E.H. Woodward in 1877 (Watson, 1906). One of the early ocher mining companies was Georgia-Peruvian Company, which was organized in 1890 and made a 50-ton shipment to England soon thereafter (Cunyus, 1933).

According to an unpublished history by Frank Dodd Smith (1983), E.P. Earle, a New York mineral broker, was on his way to Florida on the train and spotted some ocher exposed in a railroad cut near the Etowah River. He had previously imported ocher from Peru, but after finding the Georgia product superior, formed the Georgia-Peruvian Ocher Company to mine the deposits.

Unlike today's ocher mining, which is done from the surface, many early operations mined ocher underground in tunnels and drifts. After the ocher was extracted, dried and pulverized, it was shipped for use as a pigment in linoleum. Watson (1904) relates that about 75 % of

Figure 4. Ocher being processed in a rotary, vacuum filter, New Riverside Ocher Co., Cartersville, 1984.

Georgia-Peruvian's production was exported to Europe. Ocher is now being mined by the New Riverside Ochre Co., the only remaining full-time U.S. producer, which is also one of the two barite mining companies in Bartow County. Most ocher is now used to color cement and mortar.

Early reports on ocher and manganese such as Watson's in 1906 and 1908 mention the presence of barite, usually with the disclaimer that it occurred in too small a quantity to be profitably mined. However, Kesler (1950) reports a 60-ton shipment in 1895, and these negative reports apparently did not keep the Nulsen Mine near Emerson Gap from opening in 1905. This mine, the first operation of what later became Paga Mining Co., shipped some of its production to Germany, where it was made into barium chemicals (Cunyus, 1933).

Before World War I Germany had the world's most advanced chemical industry, and the U.S. chemical industry was in its infancy. It was not until World War II that the U.S. chemical industry became established in the international market.

When the U.S. was cut off from its supplies of German chemicals, the domestic chemical industry began a frenzied search for sources of supply within this country. One of these vitally needed materials was barite, and the Cartersville District, which already had a small amount of production, attracted immediate attention.

Fig. 5. Mining at Thomson-Weinman's Clayton Barite Mine prior to 1920.

Production of barite increased from less than 5,000 tons in 1914 to a peak of 111,300 tons in 1917. Although production fell to a little over 85,000 tons in 1919 when German exports again entered the U.S., companies like New Riverside Ocher, Paga Mining Co., and Thomson-Weinman & Co., which are still operating, became firmly established. Others such as Sloan and Peebles, Dertha Minerals Co., National Pigments & Chemical Co., and DuPont mined and processed barite during World War I, but later withdrew from the county.

During World War II the same circumstances arose, and barite production, which lagged during the depression, increased to a total of 1,830,000 long tons during the war years. Butts (1948) reports that nine companies produced a total of 109,000 tons of barite, valued at $929,000, during 1944. Total production to date is considered to be about six million tons. Barite is now used in more than 2,000 products, according to Stan Bearden of New Riverside Ocher Co. Major uses are in barium chemicals, in glass making and in oil well drilling muds.

Of the four states that have produced most of the nation's barite, Georgia, Nevada, Arkansas and Missouri, Georgia is the only major producer east of the Mississippi River; and Bartow County is the only county in the state where the mineral occurs in sufficient quantities to be mined.

Bauxite, the principal ore of aluminum, is a scarce commodity in the U.S. At present over 90 % of the bauxite used in this country is imported. Bauxite has been mined in Bartow County from relatively small deposits, and Georgia was the nation's leading producer until 1892 when it was replaced by Arkansas (White, 1966). The same author states that more than 40 bauxite prospects, pits and mines exist in Bartow County with production (including other Northwest Georgia counties) totaling 400,000 tons between 1889 and 1929.

Most bauxite mining in Bartow County was done by Republic Mining and Manufacturing Co., a subsidiary of Aluminum Corporation of America, between 1916 and 1926. This company exploited deposits located between Adairsville and Kingston (Cunyus, 1933).

High purity dolomite was mined by Ladd Lime and Stone Co., who converted it into a magnesium lime used to make products like baking powder and the gas to carbonate Coca-Cola. Ladd operated from the post Civil War period to the 1930s and had their plant and quarry about two miles southwest of Cartersville (Poole, 1912). Smith, 1983, states that during one period this was the largest mining industry in Bartow County.

Minerals like shale, limestone, cement rock, brick clay, graphite, sand and gravel have also figured in Bartow County's history; while others like pyrite-rich metallic mineral deposits, granitic gneiss, potash-rich slate, tripoli (a natural silica-rich polishing agent), titanium minerals, serpentine and mica may play a part in its future.

GEOLOGY

EVERYONE KNOWS what rocks are. They are those hard, heavy things that clutter up farmer's fields, get in the way of road building projects and are generally considered a nuisance rather than an asset.

Rocks are composed of mineral grains. In some cases, a rock can be mostly made of one mineral; but more often, several minerals are present with each imparting some characteristic of color, toughness and density to the rock. In addition, there are three general classes of rocks: igneous (fire-formed rocks), sedimentary (water-deposited rocks) and metamorphic rocks, which consist of those in the other two classes that have been changed, or metamorphosed, by naturally occurring conditions of heat and pressure within the earth.

Bartow County has all three classes of rocks in different parts of the county and a number of varieties of rocks in each class. Not only do these rocks occur, but they they have been jostled around by being moved up, down and sideways under the relentless pressure of colliding plates of the earth's crust. North America is moving about an inch and a half a year towards the west; however, there is the problem that all parts of the continent are not moving at the same speed and direction.

This diversity of movement causes interesting things like mountain ranges to be built, just to be worn down again; earthquakes, sometimes involving areas of hundreds of square miles; volcanic eruptions and many less traumatic events like bending layers of once horizontal rocks into folds.

The earth is not a static body. It is changing, even if by tiny amounts, every second. Since the earth began some five billion years ago much has happened. Some of the changes that occurred during about the last billion years are recorded in Bartow County's rocks - a period of time that would cover 100 million human generations, if each generation was counted as 100 years.

The earliest fossils that may be related to a man-like creature may date from about 1.5 million years ago and represent 150,000, 100-year generations. Not too bad for a species that seems determined to kill itself off; but compared to animals like the sharks that began some 270 million years ago, man has not been around very long.

Historical Geology

This is the study of rocks, how they were formed and the relations between one body of rock and its fellows. If this were a quiet planet, the rocks would be deposited in horizontal layers with the oldest rocks on the bottom of the pile and the youngest rocks on top - like a cake made of different colored layers with an icing of weathered rocks and soil on top.

In reality, all ages of rocks are not present in any one location because they were either never deposited or were washed away, or eroded, after they formed.

Recent or Holocene

These consist of the clays, sands and gravels found in creek and river valleys that were deposited over the last 10,000 years. These materials are mostly derived from the weathering of older rocks supplemented by increasingly abundant volumes of man's garbage. Millions of years from now these deposits will probably be known as the "Trash Age," if humankind is around that long.

The major features of the landscape, except where modified by man, would appear much the same to a traveler going back 10,000 years through time. A favorite fishing hole might not be developed yet, and the Etowah River might be on the other side of the valley from where it is today. But the traveler would feel very much at home and might even share a campfire with a small band of Indians.

These recent sands and gravels are now mined on Pumpkinvine Creek by the Bartow Sand Co.

Pleistocene

Commonly known as the Ice Age, the Pleistocene saw much of North America covered by thick ice sheets which reached as far south as Illinois during four successive advances of continental glaciers. The climate in Bartow County is thought to be more moderate in both extremes of winter and summer temperatures than it is now. These conditions allowed the coexistence of near tropical and sub-arctic animals such as the tapir and the spruce hen whose remains were found in the same fissure at Ladds Quarry (Holman, 1982).

Bartow County was big game country with a fauna that would rival Africa's. There were mammoths and mastodons (elephant-like creatures which stood as tall as 14 feet at the shoulder), slow moving ground sloths (long-armed vegetation eaters with eight-inch claws), bears (some of which were the size of the modern grizzly), as well as most of the small animals and reptiles seen today, along with some animals, like the jaguar and tapir, which now live in South America.

There are two localities in the county, Ladds Quarry and Kingston Cave, where large numbers of Pleistocene fossils have been found. Fossils in both locations usually represent creatures that became trapped in sinkholes or lived in the caves. Their bones did not decay because they were covered by cave deposits which preserved them.

Another type of deposit that continued to be formed during the Pleistocene and Recent are concentrations of jasperoid (hard, rusty masses of fine-grained quartz) often found near major faults, which crushed large amounts of silica-rich rocks and opened passageways for this silica to be leached and redeposited.

Outside of the changes in animal and plant life, the time traveler moving back to the beginning of the Pleistocene about a million years ago would be in familiar surroundings. He could find shelter in Kingston Cave, provided that he was willing to share it with the bears, and he could adjust to a slightly different diet.

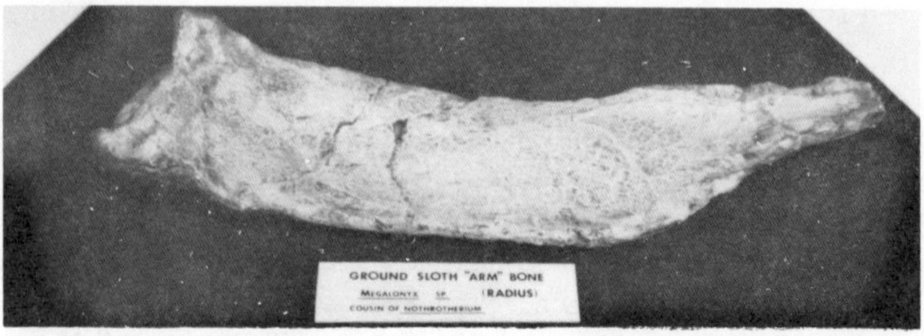

Figure 6. Ground sloth arm bone, Ladds Quarry.

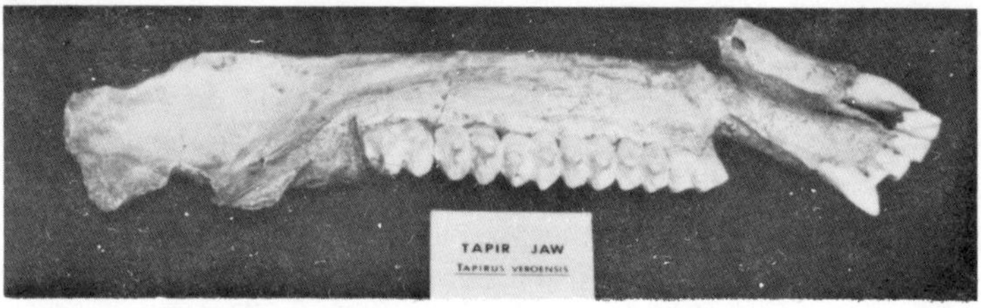

Figure 7. Tapir jaw, Ladds Quarry.

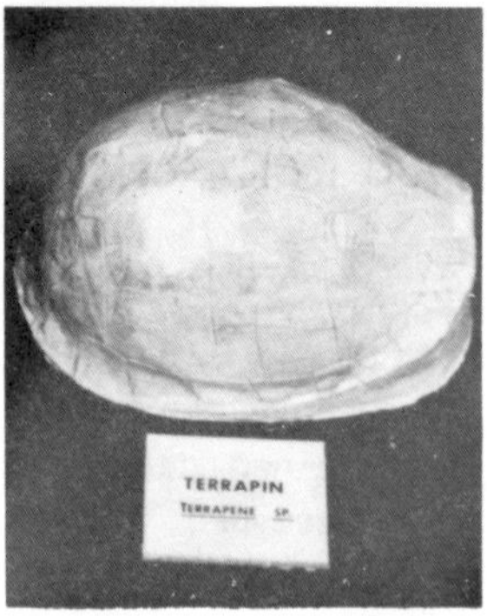

Figure 8. Terrapin, Ladds Quarry.

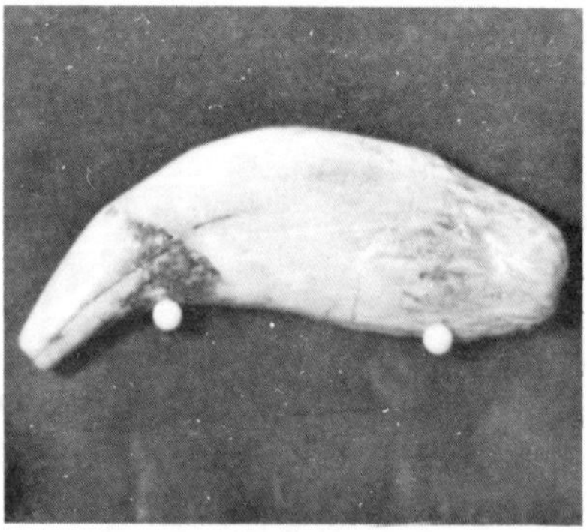

Figure 9. Bear's canine tooth, Kingston Cave.

Still, in geologic time the traveler has not gone very far back. Now he will enter a poorly known period of Bartow County's geologic history, because for the next

235 million years either no rocks were deposited in the county because it was above sea level or, if rocks did form, they have been eroded away.

Tertiary

Time is not very kindly to rocks at the earth's surface. Over millions of years even the toughest of rocks are chemically and mechanically weathered by hundreds of millions of tons of water falling as rain, carried in streams and circulating through the rocks. Weathered rocks have been found in Bartow County at depths of up to 600 feet in drill holes (Cressler, 1979).

Figure 10. The "Jug Room" at Kingston Cave.

Carbonate rocks, those made of calcium and/or magnesium carbonates (like limestone and dolomite) dissolve in acid ground waters. As the water seeps through cracks in the rocks (called joints), it widens them to the point where open cavities are formed. These cavities enlarge from pipe-size openings into caves as more and more material is removed by solution or by being worn away by sands carried by underground creeks.

Nature hates a cave, and many form and collapse beneath the surface.

Others, like Kingston Cave, may be ultimately exposed, while still more are known only as lines of sink holes where portions of their roofs have fallen. Some sinks are merely water feeders to other caves below the surface which become filled with debris.

When water carrying its dissolved load of carbonates evaporates in caves, it deposits cave formations. If water drips from the ceiling, a stalactite will form while a stalagmite may simultaneously build up from the floor. A column is formed if these two features meet, similar to the one in the "Jug Room" of Kingston Cave shown in Figure 10. If the water drips from a long crack, a delicate drapery may result.

This weathering of hundreds, perhaps thousands, of feet of rocks had the important consequence of concentrating relatively insoluble minerals like gold, barite, and iron and manganese ores near the surface. Under the force of gravity, these heavy minerals moved down slope and settled deeper into the soft clays which were the only remains of once-solid rocks.

Sometimes the clays themselves were changed. Clays washed into sink holes in the carbonate rocks were leached of their silica and converted into bauxite.

Another aluminum-rich clay which still contains some silica, kaolin, is also found in Bartow County. In Central Georgia, rich deposits of kaolin were formed near the ocean's shore where the clay-rich sediments from Georgia were being deposited during the Early Tertiary and Late Cretaceous (about 70-100 million years ago).

Tripoli (nearly pure fine-grained quartz) was also weathered from silica-rich masses within the county's dolomite units.

Flat-topped ridges at nearly the same elevation represent intervals when parts of the county had been worn down to nearly a flat surface. Kesler (1950) stated that he could identify the Highland Rim surface at between 1,000 and 1,100 feet, which he thought formed during the Cretaceous, and the lower Tertiary-age Coosa surface at about 800-900 feet.

By the beginning of the Tertiary Period 60-to-70 million years ago, the time traveler would be hopelessly lost. No landmarks would be identifiable, and even major rivers like the Etowah would not occupy their present valleys.

Some familiar animals would still be present, but many new ones like the rhinoceros-like uintatheres and the titanotheres along with some dog-size horses would give him some surprises. However, he will see even stranger life forms before his journey's done. These animals almost undoubtedly inhabited Bartow County, but because the ground they trod upon was washed towards the sea millions of years ago, none of their remains have been recovered.

Mesozoic Era (Triassic, Jurassic and Cretaceous)

One of the most fantastic group of animals ever to inhabit the earth flourished during the "Age of Dinosaurs" 70-200 million years ago. Although dinosaur fossils have recently been uncovered along the old Cretaceous shoreline near Columbus, it is not likely that any dinosaurs remains will be recovered from North Georgia.

Bartow County was an upland during the Mesozoic Era. Although it is possible that the smaller, more agile, dinosaurs inhabited this part of the state, their remains have been destroyed by weathering and erosion.

Besides the absence of mammals, the time traveler would notice a distinct change in vegetation during the Jurassic some 140 million years ago. He would still find conifers (pine-and-cedar-like trees), but the maples, oaks and hickories would not have evolved yet. Fall colors would be provided by the ginkgo trees' golden leaves.

Ginkgos were known only as fossils to the Western World until living trees were found in a Chinese temple garden. Seed were brought to the U.S., and the ginkgo once again thrives in Georgia. Not only have animal species become extinct, but many plants have also evolved and died off. Studies of fossil pollen have identified many fossil plants, and many are known only by their pollens.

Late Paleozoic (Silurian, Devonian, Mississippian, Pennsylvanian and Permian)

These are times well back in the earth's history where 350 million-year-old rocks formed during the Silurian and rocks 200 million years old were deposited during the last of the Permian Period. Whereas before, Bartow County had "sat there" with its rocks slowly weathering away, the Late Paleozoic was a time of dynamic change.

The most recent of three episodes of folding occurred during the Late Paleozoic when the Cartersville fault was deformed (Costello, McConnell and Power, 1982). This fault (a separation of the earth's crustal rocks) is one of Georgia's major structural features. The Cartersville fault and other sub-parallel faults broke up the crust to provide favorable porous sites for iron, ocher, and barite mineralization, which will be discussed more fully in connection with their rock unit hosts.

Understanding the relationships between the Cartersville fault, other faults and the rock units that they affect is still a thorny problem.

In simple terms the Cartersville fault, the Emerson fault and the Great Smoky fault are thrust faults which have brought up Precambrian and probably metamorphosed Paleozoic rocks from the Piedmont and pushed them over the Paleozoic sedimentary rocks of the Ridge and Valley Province.

As early as 1912, Maynard speculated that the Cartersville thrust fault had moved the Precambrian rocks a distance of between 12 and 15 miles horizontally and 20,000 feet (3.8 miles) vertically to where they now overlie Mississippian and older sediments.

Rife (1971) concluded from studies of fluid inclusions that the county's barite deposits (some of which filled spaces opened in the dolomite by the Cartersville fault during the thrusting episode) formed at about 5,000 feet below the surface at temperatures between 126 to 297 degrees Centigrade. He also cites T.L. Kesler's opinion that the deposits formed during the Pennsylvanian.

One of the best summaries of the area's geologic history was written by Bearce in 1982. Bearce traced a series of events which led to an uplift of the area's carbonate rocks to where the uppermost units were exposed to weathering. This uplift was followed by folding of the Precambrian and Early Paleozoic sediments. During a later, pre-Mississippian, time these rocks were refolded and sediments (such as the Ft. Payne chert) accumulated in sea-water-filled basins.

Thrust faulting, Bearce asserts, occurred during and/or after Late Mississippian time, and these thrusts were, in turn, folded by a later episode of folding. It is during, or after, this last episode of folding that the barite deposits were apparently formed, which generally agrees with Kesler's estimate of a Pennsylvanian age for the deposits.

Movement of these large volume of rocks by thrusting made mountain ranges from Alabama to Maine during what is known as the Acadian Orogeny and folded the sedimentary rocks to the west into basins. The Appalachian Orogeny at the close of the Permian formed the present Appalachian Mountains by refolding these basins.

A Mississippian rock unit known as the Fort Payne Chert is the youngest Paleozoic unit reported in Bartow County. According to Crawford and Cressler (1982), this chert unit is exposed near Emerson. The Mississippian is the lower part of the Carboniferous (coal bearing) System, when deposits of coal were formed throughout the world. Coal deposits in Dade County, which are part of the Pennsylvanian Pottsville Formation, are still mined.

These coal beds were formed by primitive plants, ferns and rushes which populated the land. During the mid-Devonian, 270 million years before, the first amphibians had crawled to shore; and it was not until the mid-

Figure 11. Coal mining on Lookout Mountain in Dade County, Georgia. At this mine the coal seam is between one and two feet thick and 12 to 15 feet of rock must be removed to extract the coal. Coal mining in Dade County was begun before the Civil War, and for many years convict labor was used to work underground mines.

Pennsylvanian, 30 million year later, that the first reptiles were seen.

The time traveler during the Carboniferous would probably seek food in the coal swamps. He would find few things on land, outside of the smaller ferns, frogs and salamanders that he would recognize. Also competing with him for food would be history's largest insects such as cockroaches up to six-feet long and dragonflies with eight-foot wingspans.

Early Paleozoic (Cambrian and Ordovician)

During this 100-million-year interval, from about 370 million to 470 million years ago, rocks which are still preserved were deposited in Bartow County. From

Table 1. Early Paleozoic Rocks

Age	Formation Name	Description	Thickness Feet
Middle Ordovician	Rockmart Slate	Medium-to-dark gray slate and silty shale and siltstone.	150
Early Ordovician	Newala Limestone	Light-to-medium gray limestone with dolomite.	200
Ordovician & Cambrian	Knox Dolomite	Dolomite with limestone beds and chert. Chert abundant in overlying soils.	3,500
Middle & Late Cambrian	Conasauga Formation	Dolomite, shale and limestone in 5 different units in central Bartow County.	3,350
Early & Middle Cambrian	Rome Formation	Gray silty shale and dolomitic shale with layers of purple, white, pale red and green siltstone and sandstone.	200
Early Cambrian	Shady Dolomite	Light-to-dark gray massive dolomite and dolomitic limestone.	400
Early Cambrian	Chilhowee Group	Upper unit conglomerate and quartzite. Lower unit graphitic slate and phyllite.	2,000
Thickness of section			9,750

Condensed from Cressler, Blanchard and Hester (1979). Geohydrology of Bartow, Cherokee, and Forsyth Counties, Georgia.

these rocks have been extracted all the barite, ocher, iron, manganese, ceramic shale, cement rock, limestone and dolomite that has ever been mined in the county. These rocks also contain most of the ground water used by the county's industries, cities and homes.

Some formation names have been changed over the years, which may tend to confound readers of the older literature. One example is the Early Cambrian Weisner Quartzite, which was formerly defined, as by Watson, 1906, as being the lowermost Cambrian formation; but is now considered a part of the Chilhowee Group. Similarly, Watson also named the overlying unit the Beaver Limestone, which is now known as the Shady Dolomite. When these name changes can be identified, they are mentioned in the discussions of the individual formations.

Rockmart Slate

The Rockmart Slate is a relatively thin, about 150-foot-thick, Middle Ordovician unit found at the foot of the Emerson (Cartersville) fault thrusts about one-half mile northeast of Mt. View Church on Route 61, south of the Cartersville Airport. It is exposed over an area of less than one-half square mile.

Described as a light-to-medium gray clay slate that weathers to a tan color by Cressler and others (1979), the Rockmart Slate is apparently more important for the geologic information it imparts than its use as a possible clay source for use in cement making as suggested by Poole (1912). Crinoids (a relative of the starfish resembling a plant) were used to date the formation.

The fact that the slate has a cleavage means that it was subjected to tectonic pressure by faulting events probably associated with the Cartersville thrust and indicates that the time of thrusting could not have been any earlier than the Middle Ordovician.

Newala Limestone

This thin limestone of Lower Ordovician age contains beds of dolomite in Bartow County with about 50 percent of the unit consisting of limestone (Cressler and others, 1979).

The limestone occurs in two lobes in the southwestern corner of Bartow County. The largest deposit underlies Taylorsville and extends to the northeast a distance of six miles, where it intersects the Etowah River. The outcrop width of this deposit ranges from less than a mile at the river to five-and-a-half miles wide at the southern border of Bartow County. A narrow band of the limestone from one-half to one-quarter mile wide parallels the strike of the Cartersville fault between Raccoon and Richland creeks.

The Newala has been used elsewhere in the state to make cement (Butts and Gildersleeve, 1948). Of perhaps more interest to Bartow Countians, this is one of the most productive aquifers in the county (Cressler and others, 1979). Fossils are more commonly found in limestones than dolomites, and the limestone portions of this unit will probably be among the county's most favorable units for collecting Ordovician fossils (See Allen and Lester, 1954, for photographs and identifications of fossils.).

Knox Dolomite

Bartow County visitors to the 1982 World's Fair in Knoxville, Tennessee, stepped on some familiar rocks, as Knoxville is the type locality of the Knox Dolomite. This rock unit is widespread and underlies much of the western half of Bartow County. The Knox is one of the county's most important water-bearing units and has also significantly contributed to the county's mineral production.

This formation consist of 3,500 feet of Cambrian-and Ordovician-age gray dolomite with thin gray limestone beds near the top of the unit. Typically the Knox is quite siliceous, yielding a cherty residue on the surface when it weathers (Cressler and others, 1979). Occasional fossils are recovered from the chert, but few fossils are

Figure 12. Part of Ladd Lime and Stone Company's processing plant near Cartersville during the 1920s. This company began mining the Knox Dolomite from Quarry Mountain soon after the Civil War. In later years the dolomite was burned into a high purity magnesium lime, which was used to make products as diverse as baking powder and the gas which gave Coca-Cola its fizz.

found in the dolomite itself, because the process of dolomitization is destructive to fossil remains.

Caves in the Knox, such as Kingston Cave and the sink-fill deposits uncovered during operations at Ladds Quarry, furnished an outstanding collection of Pleistocene fossils, some of which are shown in Figures 6 through 9. Saltpeter found in the earth of the floor of Kingston Cave was formed from the decay of the remains and droppings of countless numbers of bats that lived in the cave during and since the Pleistocene. Saltpeter mined from this cave was used to make gunpowder during the Civil War.

Fig. 13. Iron ore from Bartow County. The mass of ore shown is a secondary iron ore mostly consisting of the mineral goethite with some limonite.

The mass of ore is about one foot wide, five-inches thick and weighs over 100 pounds.

Besides being the source rock for deposits of bauxite which formed in clay-filled sinkholes in the Knox, the dolomite is also the host for some of the county's iron deposits such as those at the Hodge Iron Mine shown in Figure 3. Although all previous workers agree that these iron ore deposits are secondary (derived from the weathering of an older primary deposit) Kessler in 1950 called attention to the possibility that those deposits were the weathered residue of massive pyrite deposits.

This possibility was investigated by Hurst and Crawford in 1970, who studied 46 iron mines in Bartow County. They concluded that most of these deposits were gossans (the iron-rich residue from weathered sulfide deposits) because of pyrite found at some mines, relic textures of sulfide minerals in the gossans, the presence of higher than normal amounts of trace element zinc and to a lesser extent copper in the gossans, and the occasional recovery of chalcopyrite, galena, sphalerite, enargite and tennantite (copper, lead and zinc sulfides) from the less-weathered portions of the gossans.

Some gossans have been drill-tested by private companies who have yet to find sufficient remaining sulfides to justify production. However, the nature of sulfide deposits in carbonate rocks is that they are located in restricted and often irregular areas where porous deposition sites were caused by ancient reefs, faulting, cave collapse or changes in lithology. When the limits of these areas are reached, mineralization ends abruptly; and a drill hole missing a mineralized area by a few feet may intersect only barren rocks.

The time that primary sulfide mineralization formed is conjectural. Because some sulfides were apparently emplaced along faults and these faults would have provided conduits for mineralizing solutions, at least part of the iron and manganese mineralization in Bartow County is post faulting (during or after the Pennsylvanian).

Massive sulfide mineralization may have occurred at the same time as the barite mineralization, or there may have been more than one period of sulfide emplacement with only one associated with barite mineralization.

Besides being a host for other types of mineralization, the Knox Dolomite was mined by the Ladd Lime and Stone Company, burned into lime and shipped out in barrels (Maynard, 1912). This company began operations shortly after the Civil War and continued into the 1930s. According to Smith (1983), the Ladd operation was purchased by the Rumford Baking Powder Co. of New Jersey in the early 1900s and the high magnesium lime used to make baking powder. Perhaps their most unusual product was reported by Cunyus (1933), who states that their lime was used by Coca-Cola to make the carbon dioxide gas which gives the soft drink its fizz. Ladd also produced crushed rock, agricultural lime and, in 1920, hydraulic lime from a new plant.

Other rock quarries as well as cement and lime makers also utilized the Knox Dolomite as their raw material. All told, the Knox and the mineral deposits it contains have been very valuable assets to Bartow County, and there is a strong possibility that it will yield even more.

Conasauga Formation

Unlike the Knox Dolomite, the Middle and Late Cambrian Conasauga Formation is a "home grown" formation with its type locality being along the Conasauga River in central Gordon County. As defined by Cressler and others, 1979, the formation has five units. From youngest to oldest, these units are: 1. 300 feet of gray dolomite, 2. 700 feet of greenish, gray and purplish shale, 3. 600 feet of gray limestone, 4. 1,000 feet of greenish, gray, tan, purplish and black shale and 5. 750 feet of gray dolomite with some impure limestone southeast of Bolivar.

The Conasauga Formation is most famous for containing a natural cement rock containing the correct amounts of limestone, clay and silica so that it could be burned and ground to make cement without adding any additional materials. This deposit was was found before 1850 by The Rev. Charles Wallace Howard of Charleston, South Carolina, who started the Howard Hydraulic Cement Co. He built the settlement of Cement on the present Louisville and Nashville Railroad north of Kingston (Maynard, 1912).

Cunyus, 1933, reports that before operations were shut down in 1912, there were six dome kilns 24 feet long in use at Cement. The stack of one of the kilns remains standing as a vine-covered memorial to Howard's enterprise.

One of the shale units in the Conasauga near Adairsville was used for brick making in 1907. The brick plant was later purchased by Georgia Brick & Tile Co., who sold the facility in 1924 to a roofing tile manufacturer named the B. Muffin Hood Co. Since 1967 Universal Ceramics has made ceramic floor tiles from the shale at the site of the former Hood Co. plant. Green shale is used which, when fired, yields a red tile.

Several companies mined what was loosely called slate from the shales of the Conasauga. Shearer, 1918, describes and includes photographs of the Georgia Green Slate Co. quarry 1.5 miles northeast of Bolivar and states that later mining operations were moved to just south of White. Even as late as 1963, Croft included these units as a slate in the Conasauga Formation.

He describes the "slates" as having a good green color but were much jointed at the first mine, which made extracting the slate difficult. Flexatile, with operations near Rydal, is mentioned by Croft, 1963, as quarrying this material for use as roofing granules. This method of utilizing the "slate" was exploited by the Funkhouser Co. in 1927 to make a green-colored composition roofing (Cunyus, 1933).

These "slates" as well as the shales of the Rome Formation have a potash (potassium) content which ranges between six and nine percent (Shearer, 1918). When supplies of potash fertilizer from Germany were cut off during World War I, shales in both the Conasauga and Rome formations were examined as possible sources of potash. While it was found that potash from the crushed shales could be utilized by plants, the discovery of potash-rich salt deposits near Carlsbad, New Mexico, provided a less expensive, higher grade source.

Rome Formation

The Early-and-Middle Cambrian Rome Formation is a 100-to-400-foot thick unit consisting mostly of silty shales with some siltstones and sandstones. This unit underlies about half the city of Cartersville, and much of the valley occupied by U.S. 441 and the Louisville and Nashville Railroad northeast of Cartersville.

Compared to the nearby dolomites, the shaley portions of the formation are easily weathered and form flat areas for building, agriculture and transportation corridors. The more resistant siltstones and sandstones in the formation hold up the ridges that form the three fingers of Johnson Mountain.

The shiny shales of the Rome were considered "a soft slate" by Shearer, 1918, who called them "the Cartersville Formation," a name which has been discontinued in geologic literature.

Shady Dolomite

The 400-foot-thick Early Cambrian Shady Dolomite is the most important rock formation in Bartow County, for it hosts the majority of the county's barite, iron, manganese and ocher mines. It is typically a light-to-dark-gray massive dolomite which contains layers of dolomitic limestone (Cressler and others, 1979).

This formation is exposed in the eastern third of the county, being bounded by the Rome Shale to the west and the Weisner Quartzite and other rocks of the Chilhowee Group to the east. The Shady is commonly cut by a large number of steep faults which have displaced parts of the formation.

Kesler, 1950, and other workers have frequently noted that the Shady contains specular hematite (a red iron ore with a bright metallic luster). Kesler noted that fossils, trilobites, gastropods (snails) and brachiopods (shellfish) were most numerous in the more iron-rich zones. Maples, 1983, found archaeocyathan (an extinct primitive animal something like a sponge) and algae patch reefs in the Shady, which he postulated formed porous zones which were later occupied by barite mineralization.

Thus, the picture emerges that the Shady was deposited in a warm shallow sea some 470 million years ago. The sea was shallow enough so that sunlight the algae needed could reach the bottom, yet peaceful enough that the delicate archaeocyathans and iron-fixing algae could form colonies on the bottom. These animals precipitated the carbonate and iron from the sea water that ultimately became the Shady Dolomite.

A land mass was uplifted somewhere nearby and silts and muds were deposited over the top of the Shady, killing many of the creatures that needed clear, sunlit

Figure 14. Mining barite from the weathered residuum of the Shady Dolomite at New Riverside Ochre's mine in Cartersville in 1983.

waters to prosper. These rocks were covered by sediments that later became the Conasauga Formation, Knox Dolomite and Newala Limestone. All the while, the Shady was sinking deeper and deeper until it was buried by over 7,000 feet of sediments.

Under these conditions of pressure and the movement of magnesium-rich salty waters through the Shady, some of the calcium carbonate originally deposited in the unit became replaced with magnesium; and the Shady changed from limestone into dolomite. At the same time iron, which had been deposited by bacteria living with the algae in the reefs, became an earthy hematite.

At the start of the Taconic Orogeny in Middle Ordovician time, the Shady and overlying formations began their return to the surface and the Rockmart Slate was deposited as a shale.

During the Acadian Orogeny in Late Devonian time,

the Shady continued to be folded and pushed towards the surface. By Late Pennsylvanian time, pressures sheared the folds with regional thrust faults, pushing the Shady and other rocks back to the surface after an absence of 200 million years. These shearing pressures converted the earthy hematite in the formation into specularite.

As the Shady approached the surface from its trip from 7,000 feet below sea level to perhaps 1,000 feet above sea level, strains within the formation were causing brittle cracks to develop and stand open now that tremendous weight of the overlying rocks was reduced.

Barite mineralization was introduced after the Shady was sufficiently close to the surface that fluids could move freely through the formation and form solution cavities, because masses of barite are commonly found which were deposited in smooth-sided holes in the dolomite.

Evidence of open space filling in cracks and solution cavities points to a shallow depth of mineralization, with Rife's, 1971, estimate of formation depths of 5,000 feet being seemingly too deep to explain these feature of the deposits. At such a depth it is unlikely that open spaces would persist in carbonate rocks. The formation of at least some barite mineralization at depths of between 1,000 and 3,000 feet would appear more in line with this evidence.

If the barite deposits formed at the end of the Appalachian Orogeny at the close of the Permian, the Shady would have had about 3,000 feet of cover.

Although scattered barite is found in Polk, Floyd, Cherokee, Gordon, Murray and Whitfield counties near the trace of the Cartersville and associated faults, economic concentrations of barite have been found only in Bartow County and only in the Shady Formation (McCallie, 1926).

McCallie recognized six classes of barite deposits: 1. vein deposits (with the barite filling linear breaks in the rock), 2. replacement (where the barite replaced masses of rock), 3. breccia (where the barite or dolomite was broken up by movement along faults), 4. residual (where barite is found on the weathered surface of its host rock), 5. colluvial (where the barite released by weathering has moved down hill slopes) and 6. alluvial (where the barite is found in stream gravels).

Only the residual and colluvial deposits have proved profitable to mine, although barite was noted by McCallie in veins in the Weisner Quartzite, Rome Shale, Conasauga Shale and the Knox Dolomite as well as in the Shady Dolomite.

In the Cartersville District where the Rome Shale (referred to by McCallie as the Cartersville Shale) overlies the Shady, this unit apparently acted as a dam which served to keep the rising mineralizing solutions from escaping from the Shady Formation. Occasionally this dam developed leaks; and before the shale flowed to close these openings, solutions found their way into other formations.

The fine-grained masses of silica found in the Shady described by Crickmay (1937), from which the tripoli deposits were derived, may represent low temperature silica also carried by the mineralizing waters.

Where did these solutions come from? Probably from the huge masses of barium-rich rocks such as the Corbin Gneiss in the eastern part of the county. These rocks were also cut by the numerous thrust faults and solutions moving upward from sheared zones, which could have mobilized barium and resulted in the emplacement of barite into the Shady Dolomite and other rocks.

The source for the iron and manganese could have been the iron-rich metamorphics from the belt south of the Corbin Gneiss, which are capable of supplying the iron and sulphur to make the massive pyrite deposits which later weathered into yellow iron ores. These rocks could have also yielded the manganese, cobalt and nickel associated with the iron deposits.

There may have been several pulses of mineralization occurring within a few million years of each other with each pulse using different channelways to escape and deriving mineral-bearing solutions from different rocks.

THE MANY FACES OF BARITE

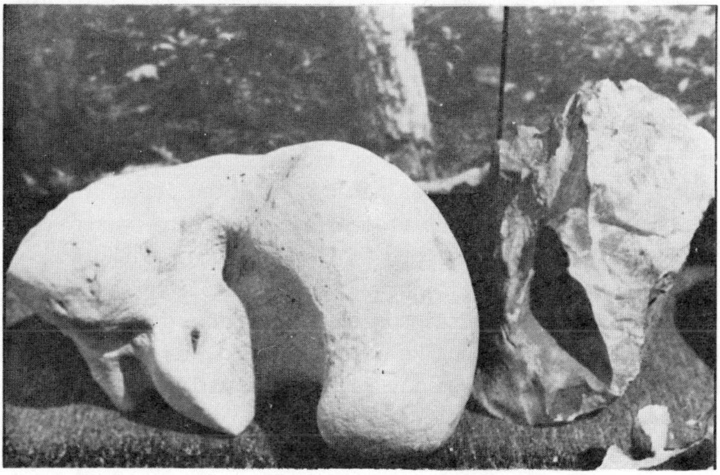

Figure 15. Left, barite which filled solution cavity in Shady Dolomite. Right, a broken piece of replacement ore.

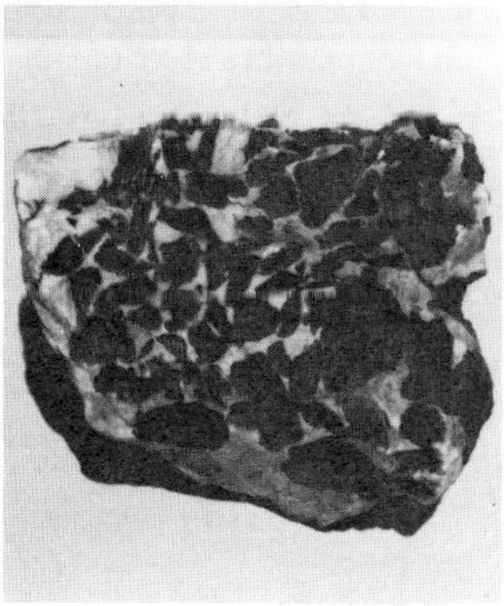

Fig. 17. Coarsely crysyalline barite with crystal blades up to a half-inch across.

Fig. 16. Barite, white, as breccia cement. Dolomite fragments once occupied the dark cavities.

If the first pulse of mineralization was iron rich and mainly from the south, iron deposits would have formed in southern Bartow County. Later, barite mineralization from the east could have emplaced barite in newly opened fractures near the junction areas of the two fault systems in the vicinity of Emerson and Cartersville.

Such a sequence of occurrences (or other theories of multiple periods of mineralization) is needed to explain why the barite does not contain the large number of pyrite inclusions that would occur if the iron and barite mineralization were emplaced at the same time.

Barite is found in Bartow County as slightly yellowish crystals, gemy transparent crystals, opaque white bladed masses and, most commonly, as white massive aggregates. Crystalline barite is most commonly found near Emerson as vein fillings. Crystals are rarely five inches long in coarse aggregates, while well-formed crystals are usually smaller than a half inch.

After being mined from the weathered Shady Formation, the barite is screened, washed, and crushed to separate it from the clay and sand. Then it may undergo magnetic separation (to remove any iron-encrusted particles), flotation (washing with a soapy solution that removes any lighter particles), and concentration using inclined tables and water.

Barite mined by Paga Mining Co. is sold to Thompson-Weinman & Co. who markets finely ground barite as a filler and extender while New Riverside Ocher's barite is sold to Chemical Products Inc., to be processed into barium chemicals. Thompson-Weinman also markets ground calcite, and Chemical Products, strontium chemicals.

Chilhowee Group

This relatively recent naming of this 2,000-foot-thick group of rocks incorporates the Weisner Quartzites as a member of the group as well as the graphitic slates and the phyllite that underlie the quartzite. The Weisner is an important marker bed for barite prospecting,

Figure 18. The contact between the Weisner Quartzite and the ocherous residue of the Shady Dolomite at a New Riverside Near Miss East ocher mine, 1984. The Weisner is the white and dark gray material in the right half of the figure, and the ochcrous contact occupies the shaded part of the figure. Manganese staining is common on the exterior of fracture blocks in the Weisner. At this location the Weisner is so weathered that it can be crushed in the hand.

since it is the lower limit of barite mineralization in the Shady Dolomite. The contact between the two formations is also where all present ocher mining occurs. Earlier operations, such at the Georgia-Peruvian underground mine near the Etowah River, also mined ocher in fault zones within the Weisner Quartzite (Watson, 1906).

Manganese and iron mineralization are also associated with the Weisner-Shady contact, and much of the manganese and iron mined in Bartow County have been extracted from this contact zone (Watson, 1908; Kesler, 1950; and others).

Flagstones have been quarried from the less weathered portions of the Weisner, such as at the Upshire Quarry near the Bartow-Cherokee county line east of Rydal. In addition, the quartzite has also been used as road aggregate (McCallie, 1936).

Graphite (natural carbon) found in amounts of from two to four percent in a graphitic schist near Emerson was first mined by Joseph F. Allison in 1892 and later by the American Graphite Co. for use as a filler in commercial fertilizers (McCallie, 1936).

This graphite was derived from the remains of living primitive organisms that inhabited the Early Cambrian seas. The time traveler visiting a Cambrian landscape 500 million years ago would have found a barren land, devoid of all life. Only in the oceans would he have encountered a group of primitive plants and animals. At one time the Cambrian was defined as the lowest fossil-containing beds, but more recently fossils, such as reef-building colonies of algae, have been recognized in Late Precambrian rocks.

Ocoee Supergroup

This group of Late Precambrian rocks has a thickness of up to 16,000 feet thick, depending on the amount of geologic section exposed between the Great Smoky fault and the older Corbin Gneiss Complex (Costello and others, 1982). These rocks occur in a north-south trending slice of rocks in the eastern one-quarter of the county and hold up such features as Signal and Pine mountains.

This unit consist of quartzite and phyllite along with coarser conglomerates. The conglomerates and quartzites support the mountains, and the softer phyllite has weathered into valleys. Costello, citing his own observations and those of other workers, states that at least part of the Ocoee rocks in Bartow County were derived from weathering of the older Corbin Gneiss Complex to the east, as fragments of Corbin rocks are found within the Ocoee sediments.

Except for minor local use for road-fill material and foundation stones for buildings, the Ocoee rocks have not been commercially utilized.

Corbin Gneiss Complex

This unit covers about 28 sq are miles in the southeastern corner of Bartow County. It consists of the Corbin Gneiss (rocks of granitic composition with flow structures), migmatite (granitic fluids injecting other rocks and converting them into a granitic composition) and some granite. Formerly known as the Corbin Granite, this unit has been more properly defined as a gneiss complex by Costello (Costello and others, 1982) and other recent investigators.

Dates taken from radiometric isotopes in the Corbin Gneiss place it as about one billion years old (McConnell and Costello, 1982). These are the oldest rocks in the county that have been dated. They formed at relatively deep depths beneath the surface, four to six miles, where temperatures were so hot and pressures were so great that the rock could flow with about the consistency of a stiff cookie dough filled with a mush of feldspar, quartz, mica and hornblende crystals.

Occasional use has been made of the Corbin Gneiss for road fill, aggregate and building stone; but no large quarries have been developed. The Corbin will probably be utilized in time to provide crushed stone to the Atlanta area market.

Limonite pseudomorphs after pyrite (limonite replacing pyrite with such exactness that the crystal shape of the original mineral is preserved) are found in cubes up to one-inch square at Red Top Mountain State Park, and rutile crystals up to about one-half inch have also been recovered from that locality.

Piedmont Metamorphics

Whereas before this geologic discussion has proceeded from the youngest to the oldest rocks in the county, it is now necessary to take a side step in time to consider these less well-known rocks.

These rocks occur in a "belt" in southeastern Bartow County that is up to 3.5 miles wide. A few of the rock types identified in this belt by Crawford (in Cressler and others, 1979) are muscovite schists, gneisses, a metamorphosed gabbro (dark heavy igneous rocks usually associated with basaltic lavas erupting from volcanoes or fissures), chlorite schists and quartzites.

A unit of metagabbro (often referred to by another rock name amphibolite after its high hornblende content) between 100 and 500 feet thick hosts Bartow County's gold deposits near Allatoona. This belt of dark, iron-rich rocks continues through Cherokee County to the Dahlonega District of Lumpkin County. By the time Yeates wrote his 1896 report, he recognized that Bartow County's gold mines were in the same belt of rocks.

Yeates (1896) describes gold workings on the Allatoona Vein in lot 929, 21st. Dist., and also in nearby placer deposits, where the gold is found in creek gravels. He and Jones (1909) also mentions the Glade Mine in lots 878 and 879 in the same district. An unsuccessful attempt was made by the Allatoona Gold Mining Company, chartered in 1931, to reopen the Allatoona Mine (Cunyus, 1933). This mine was operated during the years 1835–1840, 1878 and in 1932 according to Kesler (1950).

Within the past 15 years it has been discovered that many of Georgia's gold deposits are volcanogenic (associated with volcanic activity) and formed during island arc volcanism, although now modified by later metamorphic events (McConnell, 1980). Georgia's gold-bearing rocks now located in Bartow County were part of a group of volcanic islands with gold deposits being deposited during the waning stage of volcanic activity.

The schists, gneisses and other metamorphic rocks of the Georgia Piedmont are the most difficult of all the state's rocks to study. They are found in more or less parallel belts, forming a strong northeasterly grain across the state. Their study is complicated because the rocks are almost uniformly deeply weathered, they change along the belt because different rocks were deposited at the same time in different places within the belts and some rocks have been more thoroughly metamorphosed than others. In addition, they have been cut by a series of faults and have also been affected by the emplacement of masses of igneous rocks.

At first all of these rocks were considered to be Precambrian, but later work by a hosts of investigators has demonstrated that many of these rocks are of lower Paleozoic age. Work has now defined rocks which were originally formed by under-sea volcanic eruptions, in shallow ocean environments and by the injection of igneous masses into sediments. Detailed work has enabled a more nearly accurate picture of these past events to be drawn, but a thorough discussion is beyond the scope of this book.

Summary

Table 2 on the following page attempts to trace in tabular form the significant geological events that have occurred in Bartow County's history. The table is liberally sprinkled with question marks because there is still much to be learned about the geological processes in general and their affects on the county.

Although the earth is very old by human standards and the rocks have been much the same for the last million years or so, geology is an evolving rather than a static science. Each year brings new revelations about each of the great number of geologic sciences, and the impact of each new bit of information is felt throughout the field. In no other science is is quite so true that "The questions don't change - just the answers" because each year brings the geologist closer to an approximation of the truth.

Unresolved problems of potential economic importance to the county include the determination of the ages of the various types of primary ore deposits, their modes of emplacement and their interrelations with each other. In time, this puzzle will be solved; but not yet.

Table 2. Geologic History of Bartow County

Geologic Age	Years Before Present	Formation Name	Thickness Feet	Biological Events	Tectonic Events	Mineralization Events
RECENT OR HOLOCENE	0 to 10,000	Alluvial sands and gravels.	100		Geologically stable.	Chemical weathering of surface rocks.
PLEISTOCENE	10,000 to 1,000,000	Cave & alluvial deposits.	0-100	Ice Age mammals.	"	"
TERTIARY	1 to 60 million.	None known. Caves expanding. Coosa Plateau surface formed.		Evolution & extinction of many mammals. Modern trees.	"	"
MESOZOIC	60-200 million.	None preserved. Highland Rim surface formed.		Evolution & extinction of large reptiles.	"	"
UPPER PALEOZOIC	200-350 million.	Fort Payne Chert	50?	Many shellfish, fish, sharks. Fern-like land plants abundant.	Appalachian Orogeny.	Last stage of barite mineralization in in open fractures? Coal deposits.
LOWER PALEOZOIC	350-500 million.	Rockmart Slate Newala Limestone Knox Dolomite Conasauga Fm. Rome Fm. Shady Dolomite Chilhowee Group	150 200 3,500 3,350 200 400 2,000	Evolution of fish and sharks. Numerous shell-fish, sponges, single corals.	Acadian Orogeny. Taconic Orogeny.	Emplacement of bar-ite, iron and man-ganese ores during Acadian Orogeny?
LATE PRECAMBRIAN	500 mil-lion to 1 billion.	Ocoee Group Corbin Gneiss Complex	16,000 ?	Primitive life. Bacteria & algae.	Ocoee sediments on Corbin ig-neous rocks.	Source rocks for mineralization?

REFERENCES CITED

Allen, A.T. and Lester, J.G.,1954, Contributions to the Paleontology of Northwest Georgia: Ga. Dept. of Mines, Mining and Geology, Bull. 62, 166p.

Bearce, D.N., 1982, Lower Cambrian Metasediments of the Appalachian Valley and Ridge Province, Alabama; Possible Relationship with Adjacent Rocks of the Talladega Metamorphic Belt; in Tectonic Studies in the Talladega and Carolina Slate Belts, Southern Appalachian Orogen: Am. Geol. Soc. Sp. Pap. 191, pp. 31-45.

Bearden, Stan D.,1981, Ocher Deposits of the Cartersville Mining District: Abst. Ga. Jour. of Science, vol. 39, no. 2.

Bentley, R.D. and Fairley, W.M.,1966, The Cartersville Fault Problem: Ga. Dept. of Mines, Mining and Geology Guidebook No. 4, 38p.

Butts, Charles and Gildersleeve, Benjamin,1948, Geology and Mineral Resources of the Paleozoic Area of Georgia: Ga. Div. of Mines, Mining and Geology, 176p.

Costello, J.O., and others,1982, Geology of Late Precambrian and Early Paleozoic Rocks In and Near the Cartersville District, Georgia: Ga. Geol. Soc. 17th Annual Field Trip Guide, 40p.

Crawford, T.J. and Cressler, C.W.,1982, Talladega 'Series,' Great Smoky Fault, and Emerson Fault; Relationships in the Cartersville Area, Georgia; in Tectonic Studies in the Talladega and Carolina Slate Belts, Southern Appalachian Orogen: Geol. Soc. Am. Sp. Pap. 191, pp. 31-14.

Cressler, C.W. and others, 1979, Geohydrology of Bartow, Cherokee, and Forsyth Counties, Georgia: Ga. Geol. Survey, I.C. 50, 45 p.

Crickmay, G.W.,1937, Tripoli Deposits of Georgia: Ga. Dept. of Forestry and Geological Development, I.C. 9.

References

Croft, M.G.,1963, Geology and Ground-Water Resources of Bartow County, Georgia: U.S. Geol. Survey Water Supply Pap. 1619-FF, 32p.

Cunyus, L.J.,1933, The History of Bartow County: Formerly Cass: Southern Historical Press, Easley S.C., 403p.

Haseltine, R.H.,1924, Iron Ore Deposits of Georgia: Geol. Survey of Ga. Bull. 41, 222p.

Holman, J.A.,1982, Late Pleistocene Fossils in Georgia Jungle: Explorers Journal, vol. 60, no. 4, pp 160-163.

Hurst, V.J. and Crawford, T.J.,1970, Sulfide Deposits in the Coosa Valley Area, Georgia: Coosa Valley Area Planning and Development Commission, Rome Ga., 190p.

Jones, S.P.,1909, Second Report on the Gold Deposits of Georgia: Geol. Survey of Ga. Bull. 19, 283p.

Kesler, T.L.,1950, Geology and Mineral Deposits of the Cartersville District, Georgia: U.S. Geol. Survey Prof. Pap. 224, 97p.

Maples, C.G. and Waters, J.A.,1983, Algal-Archaeocyathan Patch Reefs from the Cartersville Mining District, Georgia: Southeastern Geology, p. 159-167.

Maynard, T.P.,1912, A Report on the Limestones and Cement Materials of North Georgia: Geol. Survey of Ga. Bull 27, 293p.

McCallie, S.W.,1926, A Preliminary Report on the Mineral Resources of Georgia: Geol. Survey of Ga. Bull. 23, 164p.

McConnell, K.I.,1980, Origin and Correlation of the Pumpkinvine Creek Formation; A New Unit in the Piedmont of Northern Georgia: Ga. Geol. Survey I.C. 52,

_____ and Costello, J.O.,1982, Relationship Between Talladega Belt Rocks and Ocoee Supergroup Rocks Near Cartersville, Georgia; <u>in</u> Tectonic Studies in the Talladega and Carolina Slate Belts, Southern Appalachian Orogen: Geol. Soc. Am. Sp. Pap. 191, pp. 19-30.

Rife, D.L.,1971, Barite Fluid Inclusions Geothermometry, Cartersville Mining District, Northwest Georgia: Economic Geology, vol. 66, pp. 1164-1167.

Shearer, H.K.,1918, Report on the Slate Deposits of Georgia: Geol. Survey of Ga. Bull. 34, 192p.

Smith, F.D.,1983, History of Mineral Operations and Exploration in Bartow County, Georgia, U.S.A.: unpb. Cartersville, Ga., 13p.

Smith, W.H., 1983, Kaolin Deposits of Central Georgia: An Introduction to Their Origin and Use: Budget Publications, Sandersville, Ga., 96 p.

Watson, T.L.,1906, A Preliminary Report on the Ocher Deposits of Georgia: Geol. Survey of Ga., 81 p.

,1908, A Preliminary Report on the Manganese Deposits of Georgia: Geol. Survey of Ga., Bull. 14, 195 p.

White, W.S. and Denson, N.M.,1966, Bauxite Deposits of Northwest Georgia, U.S. Geol. Survey Bull. 1199-M, 42p.

Yeates, W.S. and others, 1896, A Preliminary Report on a Part of the Gold District of Georgia: Geol. Survey of Ga. Bull. 4-A, 542p.

INDEX

Flexatile 27

OTHER WORKS BY THE AUTHOR

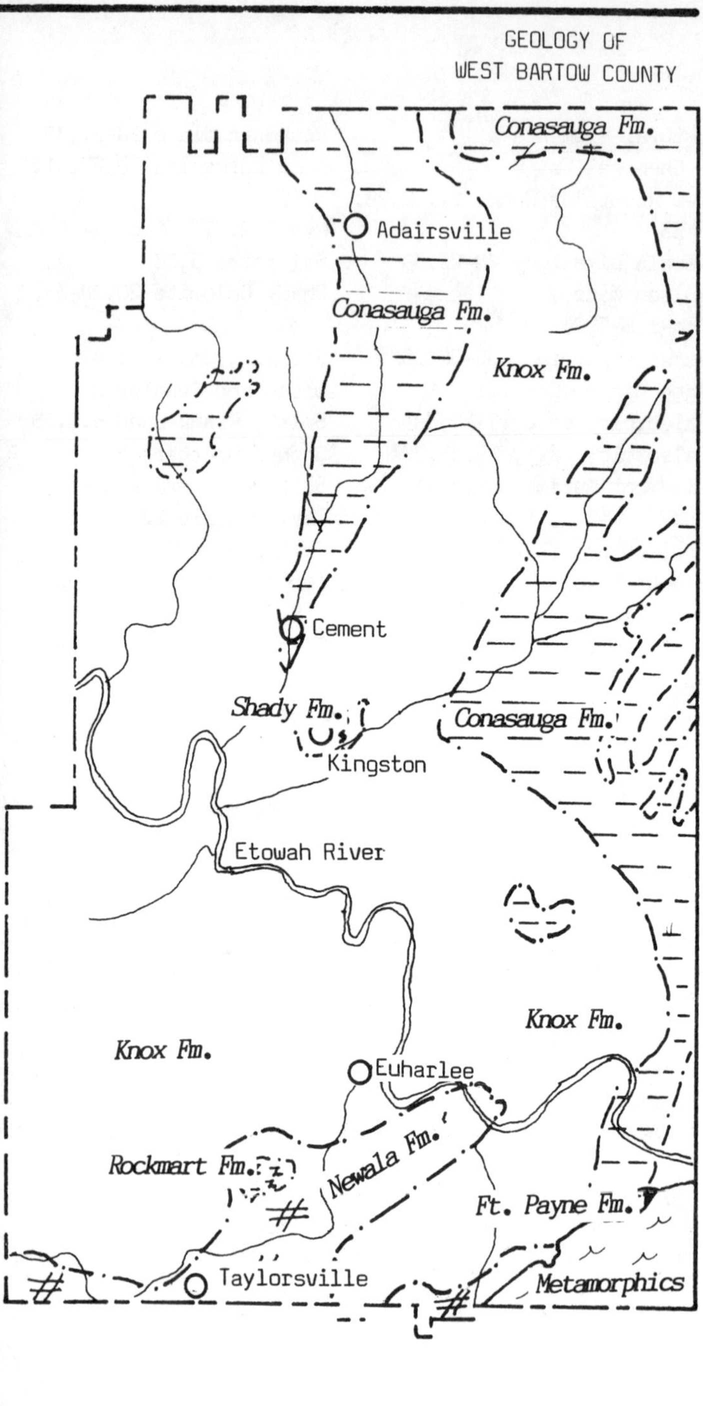

See also Table 1, page 20.